The 5 Minute Researcher

Doing better research is steps away

Sammy **Jarauvik**

DEDICATION

This book is dedicated to the researcher who sincerely seeks to become better at research – 'YOU'

ACKNOWLEDGEMENT

Firstly I would like to thank all the people who have in one way or another contributed to the contents of this book. They include students, doctorates, investigators, technicians, sales people etc

I am grateful for the opportunities I have received in research that have taught valuable lessons which are included in this book.

I cannot forget to thank my seniors and juniors who have interacted with me all through my journey in research.

I am also thankful to my wife for helping me in my efforts.

I am thankful for the support of my family behind this writing endeavour.

I have learnt a lot from my mentor and teachers during my PhD studentship; I am indebted to them for the things they have taught me in research.

I am glad that this world is filled with lots of good willed and broad minded people who show lots of kindness without which this book would not have materialized.

CONTENTS

PREFACE

Chapter 1	– What is research?
Chapter 2	– Why research?
Chapter 3	– Which area to work on?
Chapter 4	– Who will be your mentor?
Chapter 5	– Respect your mentor
Chapter 6	– Communicate with your boss
Chapter 7	– Do what you say or else 'shut up'
Chapter 8	– Maintain your lab record – 'religiously'
Chapter 9	– Mentor needs articles
Chapter 10	– Politeness with peers
Chapter 11	– Know your colleague's work
Chapter 12	– Help everyone
Chapter 13	– You gossip - you are 'gone'
Chapter 14	– Take insults lightly
Chapter 15	– No substitute for reading
Chapter 16	– Make sketches of interesting papers
Chapter 17	– Be prepared for visitors
Chapter 18	– Write to authors and investigators
Chapter 19	– Never belittle any idea
Chapter 20	– Appreciate your colleagues
Chapter 21	– Always enter the lab with a smile
Chapter 22	– Never postpone asking questions
Chapter 23	– Talk to presenters
Chapter 24	– Maintain contacts
Chapter 25	– Sticky notes do stick
Chapter 26	– Talk your work!
Chapter 27	– Yes and no!
Chapter 28	– Share ideas
Chapter 29	– Conferences and workshops
Chapter 30	– Be bold
Chapter 31	– No comparisons please
Chapter 32	– Treat sales people well
Chapter 33	– Talk warmly with lab managers
Chapter 34	– What is your homepage?
Chapter 35	– Break-time or break time
Chapter 36	– Departmental gatherings

Chapter 37	– Mind whom you are talking to
Chapter 38	– Bias in research
Chapter 39	– Never be afraid to learn new things
Chapter 40	– Never hesitate to teach science and skills
Chapter 41	– Always go neat
Chapter 42	– Machines are our friends
Chapter 43	– Back up your data
Chapter 44	– Banish insecurity
Chapter 45	– Make your presentations crisp
Chapter 46	– Know the players
Chapter 47	– Talk to your seniors
Chapter 48	– Pick up the first signal for a manuscript
Chapter 49	– Nobody likes a self-proclaimed genius
Chapter 50	– It takes time to get trusted
Chapter 51	– Treat your juniors
Chapter 52	– Extend help even beyond your lab
Chapter 53	– Keep your passport ready
Chapter 54	– Make it a goal to save 20% of your stipend
Chapter 55	– Talk to your friends and family
Chapter 56	– It's not wrong to be different
Chapter 57	– Give credits
Chapter 58	– Keep abreast of politics
Chapter 59	– Failures will come your way
Chapter 60	– Keep an inventory
Chapter 61	– Safety first
Chapter 62	– Bargaining is good; but not cheap always
Chapter 63	– Your lab is your second home
Chapter 64	– Humour in lab
Chapter 65	– Work even for non-thesis data
Chapter 66	– Find your T-max
Chapter 67	– Leave today with tomorrow in your hand
Chapter 68	– Improve your scientific communication
Chapter 69	– Reward yourself
Chapter 70	– Start a diary
Chapter 71	– Learn a new language
Chapter 72	– Remember publications by the group
Chapter 73	– Express your thoughts with your mentor
Chapter 74	– Post-doctoral position is not Holy Grail
Chapter 75	– Pay attention to details
Chapter 76	– Clarity at the end

Chapter 77	– Slip of the tongue
Chapter 78	– Pat your back and walk
Chapter 79	– Be trustworthy
Chapter 80	– Publish or get left behind
Chapter 81	– Intuition helps
Chapter 82	– Whatever you decided in the past is the best
Chapter 83	– Take care of your body
Chapter 84	– Be ready for surprises in research
Chapter 85	– I am truly sorry
Chapter 86	– Who will follow you?
Chapter 87	– Think positive
Chapter 88	– Be open to ideas
Chapter 89	– Let personal problems be personal
Chapter 90	– Value life
Chapter 91	– Not all great scientists get a Nobel
Chapter 92	– Never be intimidated by a talk
Chapter 93	– Watch your notice board
Chapter 94	– Don't grab more cups than you can drink
Chapter 95	– White board your ideas
Chapter 96	– Statistics
Chapter 97	– Software
Chapter 98	– Thesis writing
Chapter 99	– Celebrate your life

EPILOGUE

OTHER BOOK BY THE AUTHOR

PREFACE

What makes someone a successful researcher?
What makes a researcher tick?
How to be a good PhD student?
I am doing my dissertation for my master's degree, what to do?
Is research my cup of coffee?
Can I be a good research student?

These are questions loaded with intensity and these questions matter! For any researcher these are some of the myriad of questions that need to be addressed to have a satisfactory period in research.

This book is a preliminary attempt that will be useful for students undertaking research of any kind. This is a book that is intended to give a student or researcher some simple yet profound do(s) and don't(s) in undertaking research. This book is filled with 99 tiny chapters, each of which would take less than 5 minutes to read; thus the name 'The 5 Minute Researcher'

People in research, be it a small dissertation or a full time doctoral work, find time a scarce commodity and will be required to invest a lot of their time in to their work. This book respects this fact and with short chapters of less than 5 minutes of reading time researchers of any kind can give a reading without worrying about time. You can read a chapter in between your experiments e.g. incubation times. You can read one or more chapters as you wait for your glass wares to be autoclaved. You can sit down with a cup of coffee and this book during your break time. You can quickly glance through a chapter before you meet your supervisor.

I would have loved to have such a guide of simple do(s) and don't(s) before I started my research career almost a decade ago. At that time I did get some haphazard view of how to be a good researcher but none gave me a tangible list that I could refer to when required...A list that I could have looked at when I needed some guidance...A list that would have helped me keep a check on my progress. After many years in the field of research, to compile these insights in to black and white only seems more relevant; so any seeking mind in research (be it biological sciences, physical sciences, social sciences or any other field) need not fumble around trying to figure out these basic how to(s). Students can kick-start the change that they want in their research as all these points have come out of the personal experiences of many researchers.

You might not find great literary style in the chapters, but the content is needed and so has been kept simple. I have not found a common singular word to indicate masculine and feminine attributes, so I have used words like 'he', 'his', 'him' etc. It is for the ease of writing that I have resorted to the use of these words. I have also shared some of my experiences in this book just to help me articulate the thoughts in a better way; never is it indicated anywhere that I have perfected all that is written inside. I too am in the learning curve.

Though it might be tempting to infer that this book is only for students doing their research towards a PhD, it would be good to note that most of the points discussed are common to any student doing any research and related work.

This book however is not intended to guide you in your specific subjects of research, but will help you get around the common and general issues in research quicker and easier. This book will help you improve your research etiquettes. This book will tell you many things that most students are either unaware of or often overlook. There might be some points that you already know. If so – Congratulations! There might also be some newer insights which you are welcome to try. You have made a right choice to give this book a chance and it is my strong belief that you would not regret.

This will help you do a better research! Good Luck!

With best wishes,
Sammy **Jarauvik**
sammyjarauvik@yahoo.com

Chapter 1

What is research?

Well, this has to be the first question in this book. Let me give the answer that I have personally found relevant without being too long.

Research is an action, pursuit or quest by anyone intrigued by some event, phenomenon or report that he chooses to ask a series of questions and designs a doable set of experiments to find an explanation for the phenomenon or verify the report. Simply put – You see something – You don't know why – You experiment – You find the answer. That's it!

Everybody needs research at some level in his life. For example a person in a new city needs to find out the routes to his destinations. A person who builds a house has to find out the best design and architecture for the building project. A person who wants a car has to check out for the kind of car that would suit his purpose. A person who is planning to marry has to seek and find out the suitable match. All these require asking some questions and finding their answers.

Research is very much the same. You start by asking questions. You then seek to find the answers. If you are lucky, you are going to get your answers straightaway from previously published research reports. On the other hand, if you are extremely lucky you will not have a ready answer on the platter, and you will be designing a set of experiments to find the answers to your questions. You have an opportunity there. An opportunity for research!

Research is easy. It is just asking the right questions and then seeking out the answers with expert guidance, which might be through previously available scientific papers as well as through a guide or mentor.

That is all there is to research.

Chapter 2

Why research?

That is a good question.

Your answer to this question would determine how good a researcher you would be. Because the stronger the reason, the firmer would be your determination to take up research and follow it through to success. Let us take a specific example. What can be the reasons a person would want to do a PhD? I have come across many answers. Some of them are:

To get a DR before their names – Yes, prestigious to be called Doctor, right!

To get a social standing - Of course, a doctorate in a field gives you a better social reputation.

To compensate for missing out on the medical school – Oh, there are some who want to manage becoming some kind of a doctor.

Peer pressure - There are also people who say "I want to get a doctorate by research because my friends have done it"

Passion - There are people who say "I want to do research because I want to be among those who are involved in knowledge discovery. It gives me immense satisfaction and pride"

None of the above reasons can be washed off as mediocre because a reason is a reason. Period! Nevertheless, your reason will become a matter of huge importance when you get into challenges and problems in research. Only if you have a reason that is strong enough to keep you driving through your problems, will you be able to race against odd situations to go ahead and become a successful researcher. So before you start your research it would do good to sit down and do some soul searching and ask yourself. 'Why do I want to get into research?' The stronger your reasons, the better you will do in research.

Chapter 3

Which area to work on?

Without having a clear preference for an area of work (in your chosen subject) you can end up being driven by the agenda of other people. You know your field and the subjects within. So you should identify the subject that enthrals you. There is no point in choosing a subject that you detest just because the current prospects seem better for it. It is not the subject but your own passion for the subject that is going to be the driving force all through.

Let me give you an example from the field of biological sciences; the field that I belong to. I was interested in cell biology from my graduation. That really helped me choose between different areas that came my way. I was able to choose between molecular biology and cell biology. I was satisfied with my choice because I did not end up with a subject that did not enchant my heart. So listen to your heart.

May be you are into physics and you like photons or quantum mechanics. If so try to find an area to work involving these. May be you want to take up research on animal behaviour. Then go ahead and choose that area. Your chances of success are immensely higher in doing so.

There is another important thing to consider. It is understandable that many are fascinated with cutting edge technologies and most students want to work with them. But for you to excel in research you have to understand that doing a cutting edge work is not always required. Some of the greatest findings have come out of simple research methodologies.

Research as stated earlier is all about asking the right question and planning your experiments to address that question. It is not the 'cutting-edge-ness' of your work that matters but how relevant is the question your research is trying to answer. A person can work in a cutting edge field and still fail to make significant contribution to his field. So choose an area that you will be passionate about.

Chapter 4

Who will be your mentor?

There is a huge consensus on this issue that your mentor is more important than the specific topic you choose to work on. Your choice of mentor can really be an important factor deciding on your long term success in research. When you set out to look for a guide or mentor have a small check list for yourself. If not at least have a basic idea of how you want your mentor to be.

Not everybody wants the same kind of mentorship. We are all different and our needs for guidance also differ. If you think you are a person who works best under daily prompting find a mentor who has that kind of approach with his research students. Suppose you are a person who wants to work more independently, identify a mentor who is known to guide his students with significant amounts of freedom. Anything is fine as long as you are guided by a mentor of the right choice. Only you can make the right choice. And it would of course require you to meet the prospective mentor and his students.

There is one other important aspect to this. Sometimes a student with the least appealing topic can really go far ahead in research because his mentor is a person who is immensely interested in his students that he personally takes care of their student's future prospects when they finish their work. Find such a mentor and you have hit gold!

So finding the right mentor is as important as finding the right area if not more important.

Chapter 5

Respect your mentor

This is one thing that you have to follow 24/7. Any student who fails to respect his mentor will fail to inspire his cooperation. One of the secrets of being a good researcher is to show respect to your mentor by following what he says. Of course he is your boss and he requires you to follow what he asks you to do. Disobedience is one of the greatly destructive behaviours that can be seen in research students. If you want to find favour in the eyes of your mentor, make it a commitment to obey him.

Respecting your mentor cannot just be a show when he is around. You cannot always maintain a spurious show of respect. So, start genuinely respecting your mentor. Remember, even if you produce great results, the presentation of your results at conferences and their publication will depend heavily on how much you respect and hold him in esteem.

In almost all of the cases, a student who is being helped most by his mentor would be one who finds it a happy thing to respect and obey his mentor. Follow your mentor's orders without fail. Even if your mentor is giving an idea that you are sure will fail, do not disrespect him by openly disobeying him. You can try doing a preliminary experiment based on his plan and show him the results. You will also win him and his support. You will be glad for a lifetime that you respected your mentor.

Chapter 6

Communicate with your boss

Communication with your mentor is one of the pivotal factors that will help you in your research. There are certain things a mentor expects from all his students and one of the important things among those is consistent communication with him. He expects you to keep in touch with him; frequently updating him of your work progress and plans. Your communication can be in person or through mail, but has to be done on a fairly regular basis.

Failure in communication with your mentor can be severely detrimental to your research and can affect your career in the long run. If you fail to communicate with your mentor regularly he loses touch of your work. It is also possible that lack of regular communication can leave an error uncorrected and if the error is a significant one, it can end up in serious mistakes in your research.

You should share your experimental plans and keep him posted on the results you are generating. Having a regular communication with your mentor and updating him helps you to have a proper track of work progress.

I personally know the importance of having regular communication with the mentor. The many good things that will accrue will be surprising. A student who regularly updates his mentor with his work and shares the results he is obtaining would easily gain the trust and confidence of the mentor. It is not rocket science to know that this kind of student will easily find favour in the eyes of the mentor than someone who does not communicate with his boss on a regular basis.

Chapter 7

Do what you say or else 'shut up'

This one is just as simple as it reads. Just do what you say or else for goodness sake do not say you will. This not only applies with your mentor it also applies with your colleagues. People always respect a man who is a 'man of his word'. Suppose your mentor asks if you could help another student write a protocol for an experiment within the next two days and you know that you are so tightly scheduled to do justice to it, then do not commit! You can explain your situation and tell him you cannot do it and that may be some more extra days could be fine.

Once you commit to doing something and do not come up to your words, you will start losing value as a research student. In your lab if you say you will do a favour for a colleague, go ahead and keep your commitment. This would not only increase your worth in the lab but also in your own eyes. People always want to associate with a man of integrity, and integrity of commitment is of great importance.

Next time when you give your word on something, stand by your commitment, even if it means shedding hours of sleep and energy. It will pay in the long run. People will stand by your side because they know your word is as solid as rock.

Chapter 8

Maintain your lab record – 'religiously'

Now, this one is not as easy as it reads. Superficially this looks so easy. But the truth is; this is one of the frequently overlooked necessities by many researchers. Many students fail to commit themselves to maintaining their lab record. This one task is going to be very boring at times and you will be tempted at the end of a tiring day to go to your room, take some rest and update your work in your record book later. But I urge you do not cave in to it. That can become very damaging. Never leave the lab without filling your record book.

One of the questions that research students have is– "what do I record in my note?"

Record everything you did in your record notebook; from the protocol you followed, the reagents used, the stock used, the modifications you used...anything you did. Some things might not make sense while you are writing but at the end of your research period when you turn back and have a look it would make a great contribution.

You should honestly write every single thing that you have done with the experiments. You cannot bank on your memory to remember all the small details.

Why unnecessarily tax your brain with details that you can simply record in your record note and carry on with your more important work or experiments?

Chapter 9

Mentor needs articles

This is one aspect that would help every research student. Your mentor, though more expert in the field, will obviously be busy with hundred things. He would be busy taking classes if it is an educational and research institute. He will be busy attending meetings. He will be occupied with grants and research committees. He can be involved in many other panels which would require much of his time. So it is very much understandable that your mentor needs a steady supply of good research articles to keep him primed with the latest in the field (however this does not mean he will be lacking otherwise).

If you come across some interesting article (you better, ha ha ha!) send it to him with a mention of the salient points. The research article need not always be directly connected to your work; it can just be something that has a wonderful logic or a neatly organized experimental process.

Apart from keeping him updated, your endeavour would also help him know that you are serious about your work and that you are not just spending your time in the lab casually. He will come to know that you are interested in being productive. Moreover the very act of choosing to send a research paper to your boss would keep you focused on the scientific content; you will end up reading that paper from top to bottom. Such exercises accumulated over a period of time would certainly enhance your scientific acuity. One day this would have its great reward.

Chapter 10

Politeness with peers

Do you want to be in the presence of rude and arrogant peers? Do you want to associate with colleagues who don't care how they behave with you? I know your answers.

This chapter is not about changing difficult peers but changing ourselves. If you want to have a peaceful existence in the lab being polite with our peers is vital and indispensable. If you are a research student, on an average, you would be spending more than half of your waking hours in your research environment. Now that is a huge amount of time. I would better be very careful how I deal with my peers.

They are the people who are going to help us. Don't ever think, "Oh, why should I be good to that guy, I don't share a small pipette with him". No, that is not how research works! Research is as much people art as it is hard-core subject. Being polite is certainly important. When a colleague comes and asks you in the midst of your experiment for a catalogue number of a reagent or for some information, you should know better than mauling his sense of timing. Instead you can and should politely give back a genial answer. It helps in your research because you cannot see the face of a person you hurt yesterday without having to cringe, and that would affect your thought process the next day. So make it your commitment to be polite to your peers.

Chapter 11

Know you colleague's work

Phew, this might seem uninteresting to many. Someone might ask. "Why on earth should I know about my colleague's work?"

Yes, it helps. This might not be appealing to some. This can be difficult, especially when your colleague is working on a tangentially opposite topic. But it pays to know what each of your colleague is doing.

There are many advantages of keeping yourself aware of what your fellow players are doing. There could be something you can learn from what they are doing. It may just be an insight. It might be the way they design their experiments. It can also be the way they plan their work which if adapted could improve your performance too. You could also get some research idea and your discussion with your colleagues might just be the spark needed. Getting to know what your colleagues are involved in would also indirectly help you grow in your people skills and develop your potential to be a team player. When they find that you have a healthy interest in their work, they will also be willing and ready to share ideas and suggestions about your work which might prove valuable.

So never hesitate to know about your colleague's work.

Chapter 12

Help everyone.

This is one area where people tend to be very selective. This is one virtue that could take somebody far and long not only in research but in any vocation. What is meant by helping somebody? It means extending your support either through your physical involvement or through your scientific contribution towards their research.

When somebody comes to you and asks for a help which you could do, go ahead and help him without giving a thought about how you would benefit. Your help should never be dictated by your desire to get a help in return. Helping others, with a genuine goodwill, would make you a great asset in you lab and you would be a sought after person.

Remember, a person who helps others with a sincere heart is always in demand!

Chapter 13

You gossip - you are 'gone'

This is very important. Stories could be written on how people who were progressing were brought down because of their tendency to gossip. Gossip is simply saying something about another person that is untrue or unneeded. Students who indulge in this behaviour often have some people surrounding them because of the cheesy stories they propagate. This could serve to boost their sense of importance. But this has to be avoided if you want to be a better researcher.

When you gossip, invariably you are character assassinating; damaging another person's image. This will have unpleasant effects. Rest assured, one way or another, the thing you spoke will reach the person you spoke about and you will be losing that person's cooperation soon. Research is all about cooperation.

There is another side effect. The person with whom you are gossiping will lose his trust in you because at the base of his mind he will be contemplating when and where will you be gossiping about him. So you will be losing the respect of the person to whom you are gossiping as well. Thus you are building walls around yourself in research, wherein the levels of cooperation will come down. So avoid this detrimental behaviour.

Chapter 14

Take insults lightly

Researchers need to learn to take insults lightly. This can dramatically improve their productivity. No one is immune to insults. We all, at one time or other, are taken by surprise due to the insults hurled at us. In order to be effective in research you need to be aware that insults, if taken seriously and in to our mind, can start blocking our effectiveness. You cannot always shut up the people who insult but you can decide where you place their insults.

When someone insults you, your work or the data you have generated, the best response would be to thank him for his inputs without sounding condescending or resentful. Just thank them and move on. Don't take their insults seriously. I know a guy who was told that he was not fit for PhD. I know of another person who was told that he was too old for research. But the best thing you can do to yourself in such a situation is to keep moving.

Now, there is difference between constructive criticism and insult. A constructive criticism serves to give you a better perspective and an insult is just to make you feel bad. Take such constructive criticisms well but insults are not worth your precious time.

Take insults lightly and soar higher because only your peace of mind is one of the strongest factors for doing better work. Rest assured when you take insults lightly you are growing!

Chapter 15

No substitute for reading

If all the points in this book have to be reduced to ten, this one point would certainly be among them. How many research articles a student reads over a period of time would certainly be a great deciding factor on his success level. Trust me, you would become really effective in your lab and the results would be very apparent as your scientific reading increases. Your mentor would see the change in you soon. Your own confidence would increase. Your own perspective about research would be elevated several notches.

Many students just slip away from a crowd that is discussing subject just because they are not well informed of what is happening in their field. Only if a student knows his field he can boldly stand in a group and feel good about it.

There are different ways to approach this. One way is to start by making a commitment to read one research article a day from the title to the conclusion. Now, if you are someone who manages to read many articles, it is fine. But if you are someone who is confused about what to read, you should choose one article that interests you and read it with a pencil or pen in hand to take notes.

You should determine in your mind – "Come hell or war, I will read a minimum of one research article a day"

Over time you can increase the quantity. Just imagine how much more knowledgeable you will be at the end of just two months. Your knowledge will come to your help at the right moment.

Chapter 16

Make sketches of interesting papers

If you have read an interesting article then it would be good for you to make a rough sketch of the findings and the experiments described in it on a paper. It would be worthwhile to have a small note book for this purpose. I had one called 'LIT-BIO' meaning literature in biology. It was amazingly useful when re-checking and revising important research publications.

It would be waste of time to always rummage through a heap of print-outs to find that one particular article which you desperately need. When you jot down some ten or twenty salient points about a paper with its reference in your notebook you also would have a pictorial memory of the points for easy recollection later.

Similarly it would be worthwhile to make presentation slides of few of those interesting papers. This would come handy if some sudden journal presentation arises, which are not uncommon in the research set up.

Chapter 17

Be prepared for visitors

It would be a great idea to have a presentation ready about your work in your laptop or lab computer. It need not be your data always. It can be the basic work plan you have and some of the designs of experiments that you follow. To have such a presentation ready would really put you and your boss at ease in the event of sudden visitors, which is common in research. The presentation need not be a two hundred slides showdown; it can be just two to four slide consolidation of your work.

So when some eminent scientist from across the country or globe walks in to the lab and your mentor asks you to explain your work to the guest, you will glide through it (later you will be hugging yourself that you had this presentation ready). You would give a great impression about yourself and your boss. Who knows the man to whom you present your work might one day be your next mentor? The world of research is very small!

Chapter 18

Write to authors and investigators

This is a rarely found practice among research students. Researchers are very busy that writing seems to be a trait that is reserved only for grants and manuscripts. We all generally tend to request authors for the full text of their research publications and thank them. But if an article is freely available, there are only a few students who make it a point to write back to the authors and the lab from which the work originated, expressing thankfulness and appreciation for their publication.

This is a small gesture that can fetch you greater goods in research. You would also have more contacts with other scientists around this planet. This is one sure way of establishing new contacts if you are seriously interested in widening your influence in research. This can even help you carve your future niche as these are important people around.

Chapter 19

Never belittle any idea

Ideas abound in this world and more times than one could care to calculate has an idea that was initially rejected, become a world changing phenomenon. So as a person in research you have to be very careful to never belittle any idea. If someone comes up with an idea, however silly or simple it might be, it would be good not to belittle the idea or the person. Ideas evolve; today an idea might be small or silly but tomorrow the same idea can give rise to a glorious idea that can change the paradigm of understanding in research.

Moreover when you belittle someone's idea you end up alienating them and this would serve to close the doors of cooperation and goodwill. What one has to realize is that an idea is not what makes it great but the mind behind. Sometimes by discouraging an idea one can even end up mauling the enthusiasm of another person. So this has a potential to be a hampering factor in research. After all isn't research a vocation where different ideas are to be put out and tried?

If a person can come up with an idea, however silly it is today, he can come up with a better one next time. So belittling an idea is one of the worst things that anyone could do in research and this has to be avoided at all costs.

Chapter 20

Appreciate your colleagues

Now this one is easy and tough. Giving appreciation is not something that all of us can spontaneously do to everyone we interact with but with a disciplined and determined mind we can. Why should one appreciate others in research?

The answer: appreciating your colleagues would always have a positive ripple effect on you, provided you do it genuinely. You have to be true when you appreciate somebody. Who will ever want to interact with someone who knocks the door only when there is a need and is gone when his need is met? We are all social beings and very much in need of this tonic called appreciation.

When we appreciate somebody honestly, we are in effect creating a healthy atmosphere where friendliness flourishes. Only when your work place is filled with good-willed people, will you have the needed motivation and strength to push forward. The world of research is small and who knows, one day you might meet or even be required to collaborate with one of your present colleagues. It's always good to build your relationship with them and appreciation is one way.

Chapter 21

Always enter the lab with a smile

This is one thing that you can put to use immediately when you reach your lab. Smile is an inexpensive way to boost your looks. When you enter the lab with a smile you set the mood for the day. When people see you smiling and entering the lab there is a positive attitude that is generated.

Have you experienced the sulking presence of a grump who is serious and frowning? If yes, then you know what kind of effect you will create if put up a sourpuss look in your research environment. Enter your lab with a smile and greet your lab mates warmly every morning. Remember, you would be interacting and talking with them more than you would be doing with your social friends. And this interaction will have a strong impact on your research because, if you feel out of place in your work, you will not be in the best of your abilities in your work.

Like it or not, your colleagues and peers are the ones who would be in your circle of influence at the workplace. They would have a bearing on your work. Getting to work with a smile always elevates you and others. This gives you an energetic start which serves to overcome the morning inertia, especially when you start checking your to-do list and are looking at the great list of responsibilities. Smile is a great way to have a great day! Adorn yourself with a smile and start your day with positive energy. You will not only attract friends but enhance your efficiency.

Chapter 22

Never postpone asking questions

This is one of the negative behaviours that can be seen in research: A student having a legitimate question and wanting to ask but not asking for the fear of being ridiculed. Each one of us has times when we have this dilemma of whether to ask a question or not, especially in a forum where our colleagues and superiors are present. You might encounter a situation where your head is throbbing with a question and you are hesitating. Don't think – 'what if I look stupid' or 'what if people laugh'. Just remember this - How many times have you walked out of an auditorium or hall thinking, 'I should have asked that question and not worried about others'?

Everybody wants to maintain their reputation and so do you. Such experiences are common. These are fears that all research students battle with. I have experienced many such fears. But to be a successful researcher, one should not let such inhibitions stop him from asking questions.

One of the best ways to push yourself to asking your questions is by raising your arms as soon as there is a call for questions. Once you raise your arm there is no going back. Just don't think even for a second more. Go, jump into your question and ask it. Be it a seminar or a simple science discussion over a coffee table; take an oath never to postpone a question for which you truly do not know the answer. You might be laughed at for a few seconds, but remember, people don't think about you as often as you think they do. Go ahead - let people laugh, but eventually you learn. You will be wiser at the end of the question. Bolder too!

Chapter 23

Talk to presenters

This is a principle where you can positively influence your peers in research. Any research student will be exposed to lectures and presentations by his colleagues. These can be anything from a scientific paper presentation to the weekly laboratory research updates.

It is always good for you to wait and congratulate the speaker, after the presentation, sharing a few friendly words with him. You can tell him what was most striking in his talk. You can discuss some of the inferences he made. You can ask him a question or two if you don't understand something (and that will happen if you do not apply the previous chapter, ha ha).

This will not only give you a better understanding on what was presented but also deepen your rapport with the presenter. This behaviour helps you to get accommodated better in the sphere of research. You will find a sweet spot within a presenter when you frankly discuss about his presentation. He will be immensely impressed and be interested in your work. Try this in the next talk and you will be surprised. Not only will you gain people skills, you will also end up more knowledgeable when you have a one-on-one chat with someone about a topic.

Chapter 24

Maintain contacts

There is this funny story of a car thief who was arrested by the local police and taken to the sheriff. He was finally presented in the court. The opponent's attorney waged a severe legal battle and everybody in the courtroom was sure the thief will be convicted. But the judge finally gave the verdict – 'not guilty'. Everyone was surprised. The fact was that the thief and the judge were close childhood friends while in their primary grade. The moral of this story: Contacts are important.

Humour apart, in research it is always important for us to build our network. This is of great importance if you are a budding student wishing to pursue longer time in research. If you are doing a PhD you would certainly need to have contacts to further your plans in research. It is good to develop and maintain such contacts.

One of the ways in which you can do this is through scientific meetings. When you go to workshops and conferences, you will get chances to meet lots of scientists who work in your area or similar areas. You can grab such opportunities to make contacts. There is no need to think or be diffident in approaching established scientists. Please make sure that you talk and discuss your work with these kinds of people when you meet in such gathering. You never know, one of them might be associating with you in the coming days. Never underestimate the power of making and maintaining new contacts.

Chapter 25

Sticky notes do stick

This is one formula that most of the management gurus would recommend to anyone interested in having a productive day. This is applicable to research as well. When you enter your lab take a piece of paper, a sticky note or a plain diary and list the tasks that you would like to finish by the end of the day. Just keep writing everything from the smallest blade to the biggest boulder that you want to move that day. There is no need to be selective about what to write; anything that needs to be done needs to be written down.

Then start by picking up one task at a time and complete it. When you finish one task, strike that point with a thick dash. At the end of the day you will find that you would have done more things than you would have imagined possible. This is because you have removed the clutter from the mind and put it on paper.

The best thing about this plan is that as you keep progressing through the day, the number of items with the thick dash would keep increasing, which by itself would give you the drive to finish the list. It doesn't matter even if you don't finish the entire list. All that matters is whether you set out to start the day with the work list.

Importantly, make it a point to take down the tougher tasks first. It is easier to finish the list with easier things at the end. If you try this you will never doubt its effectiveness.

Chapter 26

Talk your work!

I have personally seen people who, when asked about their work, say that they are just working on some simple project of which there is no real value. *Many students tend to think that giving out a great impression about the work they do is equivalent to boasting.*

The image such students create by holding on to such behaviour is that they are unimportant in the world of research and that their research is insignificant. If you think you are exhibiting such behaviour, you can very well stop doing that and stop it for good. Always be proud of what you do. It doesn't matter if you are working on the god particle or road banking. Neither does it matter if you are working on embryonic stem cells or epidemiology. You should be proud of your work, because any work is great or small only by the researcher who is undertaking it.

So the next time somebody asks you about your work, talk about your work with a healthy pride. Tell them you are working on an exciting topic and are enthusiastic about the goals you propose to reach. If you are not going to talk positively about your work, no one else will. People generally move away from you with a sense of negative feeling if you are not talking boldly and proudly about your area of work. Talking positively and with pride would not only give a good impression about you, you yourself will feel much more confident if you talk good about your work.

Chapter 27

Yes and no!

This is another area where lot of research students falter. It is quite common and if you think this describes you, there is no reason to be surprised as this is one of the widely acknowledged problems of research students. This is an issue of how a research student answers questions.

A student has to know how to answer people who ask questions about his work and the data that he has generated. When you are faced with a question, and are sure of an answer, then you should reply with an emphatic 'yes' or else give a clear 'no'. I have seen many students getting intimidated by the question or the questioner that they end up fumbling with their answers. Not giving a clear answer would result in people making assumptions about you and your research. And such a thing might not be good. You should always have a clear conviction about your work so that you are able to answer questions that come challenging you.

Of course, there might be exceptions where a clear yes or no would not be possible. As much as possible be clear in your answers. A clear mind is always respected!

One of the secrets to be effective in research is to be clear when there is reason enough to be clear. If you are not clear, then make it a point to be clear that you express your lack of clarity and get back to that question with a clear answer in time. People always appreciate honesty in research.

Chapter 28

Share ideas

Sharing ideas is one of the keys in having a successful research career. This is an area where there is a huge heterogeneity in opinions. There are groups which are for and against sharing ideas. I would strongly suggest that you talk about your ideas with your mentor, seniors and colleagues.

There is no need to feel afraid or ashamed. There is a great deal of intellectual property issues that are going on, that people are paranoid about sharing ideas. Plagiarism is on the rise making people little wary of letting their ideas open to others. It is understandable that there are situations where you just do not open your mouth and give away your plans in your research. But, this need not be the case with everyone. You can certainly share it with people whom you trust and who can handle the sensitivity of your ideas.

Don't think about keeping your ideas secret for the fear that anybody and everybody might steal it, and mint money out of it. Sometimes it would be good to gauge the genuineness of the recipient and take the daring step of sharing your ideas with them. Without this risk you would lose many valuable suggestions and corrections. So remember, it is you who has come up with the idea and if you know 'what' then you can work it out faster than any other person. So don't worry about sharing them with dependable people.

I heard a great professor say when addressing such an issue, "If you can come up with one idea then you can come up with ten others as well". So share your ideas with worthy people.

Chapter 29

Conferences and workshops

This is an area where every research student has to keep himself updated and informed. You should be up to date about the various conferences and workshops in your field. It is good to keep yourself abreast of the latest information regarding such wonderful opportunities to show case your data and talent. These kinds of gatherings also give you unparalleled ways to connect with the other players in the field. These kinds of gatherings can also open up newer ways to get expert opinion about your data.

So make it a point to search the web at least once a week for national and international conferences. If you are a PhD student, it is of paramount importance that you attend a minimum of one national and one international conference to present your work. If you can make it to more of such scientific meets, it would be better. Thus make it a strong reminder that you have to keep abreast of the scientific meetings and gatherings happening around the world. You would never regret for knowing more.

Chapter 30

Be bold

I personally have experienced and seen many other research students having a timid attitude. This need not be the norm, but every student has periods when his courage is tested in research.

Boldness is one criterion that is essential for you to be a great researcher. There is nothing to fear in research as long as you are doing your duties. Do your duties and remain fearless. Even if you have a tough time with your work or the people around you, you only need to gear up for facing the challenges and meeting the deadlines in relation to your work.

Never fear! Do you work and let the world fall on you and break. This need not be restricted to your laboratory work. I personally know of students who keep mulling over the thought of asking for a day of leave. Suppose you want a week off from work, stop worrying about it. Go, tell your boss that you are stressed and need a break. Or else, you might need a new set of reagents or apparatus, go to your supervisor and request him.

Be courageous! Courageous people go far in research.

Chapter 31

No comparisons please

Oh, the woe of comparison that seems to taunt many in research! Of course, this is a competitive and comparative world. That does not mean every student has to measure up his worth by comparing himself to someone else. This bane called comparison is rampant in research.

Many research students weigh their worth by comparing themselves with other researchers. You already know that no two human beings are completely alike and so attests science (not even monozygotic twins; they have differences). So quit trying to see if you match up to that senior who has got twenty publications during his doctoral work. You are never going to be him and neither is he ever going to be you. Don't try to compare yourself to your school mate who finished his doctorate so early that you end up pressurizing yourself.

The best way to overcome this is to try and compare yourself with what you were yesterday or last month. That is a more reasonable scale of monitoring progress. Won't you like to find yourself doing a better job today than you were doing last week or last month? Comparing yourself with others will only serve to increase the feelings of inadequacy and push up the levels of stress. You already will have enough tension with your research. Why would you want more?

Quit comparing yourself with others!

Chapter 32

Treat sales people well

Aha, this is one area which requires reiteration. Sales people form an integral part in the life of a research student. They are the ones who bring in vital information of a new product or technology that might be beneficial for your research. How you deal with them will tell a lot about how easy your progress will be in your field.

There will be many situations when sales people from different companies catering to your experimental needs, will come knocking your lab door. With loads of work (and the lots of research papers to read), it would not be surprising if you find yourself a little edgy when a sales person comes knocking. But, hold your nerves and remind yourself that they are there to help you in your research.

Even if you don't feel like talking with them just manage to spend a few minutes and wind up the conversation. That would be any day better than showing up a face. There are many advantages of having a good rapport with sales people.

I have personally have received huge benefits by treating such people well. In one instance, I got a whole vial of a costly antibody free of cost through the sales person just because he felt welcomed in my lab. In another instance, I got the help from a saleswoman who went out of her routine day work to get the complete price list for more than a hundred laboratory reagents and plastics (for writing a research grant). There are many more. But I just want to tell that I got all this just by treating them nice. So without doubt it is of huge importance to treat sales people well.

Chapter 33

Talk warmly with lab managers

When you are in research, you are not alone. You are not an island. You would obviously be requiring the help and cooperation of other staff. How many researchers do you think take the effort to smile at the security staff or the secretary in the office? All these people indirectly contribute to your research and ultimately your well being.

There are many such people who, without their knowledge, play a crucial role in your research. Even the person who cleans the floors in your laboratory daily is important for providing you a clean workspace. It might be the technician who washes your used equipments and reagent bottles, giving you a clean start every morning. It is important that you treat them warmly. A gentle wish or a caring enquiry can really put them in high spirits. Everybody needs to be treated with a sense of concern for they too are humans with problems like you and me. A gentle treatment would encourage them to provide their service even more happily and effectively. Ultimately you will be the beneficiary.

Chapter 34

What is your homepage?

Today is the day of social networking, and websites for these purposes abound. It is very important as a researcher to keep your focus on your work during your working hours. If yours is a personal computer, this idea will be easy to implement. What you have as your homepage in your browser will definitely decide your scientific mood as you start your day. For instance if you have a social networking site as your homepage, chances are that you will invest the first energetic minutes or hours of the working day on your personal interests after you enter your work place. This can delay your work.

If you make a scientific journal or scientific search engine as your homepage, as soon as you hit the browser first thing in the morning, you will have a stronger inclination to stick to your research and not dilute your productive day with distractants. Now, I am not against social networking. I am just advocating judicious use of time. This will certainly set the mood for the day. It is also good to have a separate email account for official work and refrain from non-official browsing during working hours. This will certainly increase your productivity.

Chapter 35

Break-time or break time

Break times are very important for any work and more so for research. Human mind cannot continuously work without having a time for relaxation and rejuvenation. How you handle your break times will tell a lot about your style of working. Coming straight to the point: When taking a break from your work, have a determined mind to only take the time needed for your break. Do not overshoot the time.

For example, if you go for a coffee break, take only the time needed for a coffee and come back, if not tell your senior or supervisor that you are going off for a longer time to get a breather. They will respect you for your integrity than finding you near the foyer chatting about the latest craze in Hollywood. This will not only improve your working efficiency, it will give you a healthy self-respect.

People will always respect a person who sticks by his timings. I have both failed and succeeded at this. There have been times when an interesting talk over a cup of coffee took a lot of time that I was forced to postpone my experiments. There have also been times when a regulated schedule during break times resulted in getting more done. So keep a check over your break times.

Chapter 36

Departmental gatherings

Now, this is one lighter point that is worth mentioning. Every department has gatherings and meetings (non-academic). These meetings might range from a small birthday party lasting 10 minutes to a longer evening when your colleagues go Dutch. Sometimes even your mentor might take his students out for a coffee. These are some times that you should try to part of. Of course, unavoidable situations may prevent you at times, but for the most part it is important that you take part in such social gatherings.

It is in such casual gatherings that you sometimes get to know your colleagues well. These can be times when you really get to have a closer talk with your mentor. It also creates togetherness in a team which is very important for research. It is in such social gatherings that you can build stronger rapport and contacts within your team. If your lab is planning for a tour, volunteer for help. Suppose your department is organizing a New Year party, take pro-active part in it. You can even find your potential spouse in such gatherings (Now, that was just for a laugh, okay. Please don't go spouse hunting on such occasions).

Chapter 37

Mind whom you are talking to

This is just a piece of information that I have experienced and seen. There are times when you just cannot keep talking your research. Period! Though one might like it, it is still better to gauge the kind of audience or group we have before opening our mouth.

If you think the group you are in is one that is more casual and non scientific, try talking something non-academic that will be of interest to the group. You will end up like a dinosaur if you start talking about science and research (People might not relate to you). Do not be overly 'researchy' during such times.

At the same time if the crowd is scientific and you don't have anything in science to talk about, you will be an outcast. You should always be ready to ready to talk science when required.

So mind your audience and do the talking.

Chapter 38

Bias in research

This is a very important aspect that has to be kept in mind. We all have to respect and treat people of the opposite gender with dignity. Research is a vocation where only minds play the different roles. There have been times when a person was being discriminated just because of the gender. But you and I are called to take responsible role towards this.

It is important to have a mindset that allows and calls for equal treatment of both the sexes. Above all, everybody deserves to be treated equally and there is no single factor based on which we should treat fellow researchers unfittingly. Oh you are saying, 'Well, I find such things happening and what can I do about it?". You can be the change that you want. You be the example.

Similarly it's time we got beyond the colour of the skin. Having a judgement and notion about a person based on his race and skin colour is not just bad; it is unfitting of a human being. A researcher finds new things for the world. He should not be stuck up with old ways that are compromising to the quality of human life.

Chapter 39

Never be afraid to learn new things

Many research students have this problem. While in research, suddenly you might be expected to do a different experiment or undertake a study design that requires you to learn something new. It can be anything; a new technique or methodology. Never hesitate or be afraid to learn.

Let me share a small experience here. During my research I had to do some immunofluorescence work (if you are from a non-biological background and don't understand what it is, never mind). I was diffident because I had never done it in my life and was fidgety whenever my boss talked about it. I kept postponing it until one day when I could no longer escape the necessity. So I decided to take the help of a kind senior. It just took me a less than a week to learn the technique and produce results. I was surprised at my own fears that had held me back for months from learning a technique which was learnable in a week.

Now, this might seem like an isolated incident. No. I personally know other researchers who get a little unnerved at the thought of learning and doing something new. In my case what I feared, I could do later without any supervision. That is exactly the same with any technique. Only yours might be easier to learn or take a longer time.

So never be afraid just because you don't know the technique. You will learn it once you put your feet into the waters. And once you learn it will become so simple that you will be amazed at how easy learning new things in research can be.

Chapter 40

Never hesitate to teach science and skills

Life is a race according to many. Even research is being undertaken by many as a race. People compete and want to outrun others working in the same area. This also breeds a kind of fear; fear of sharing science and skills. It is not uncommon to find people in research who are wary of sharing what they have already learnt with their colleagues and associates. Their reasoning: why should I teach something to somebody that I learnt the hard way? Now, my friend, if all the scientists and leaders thought that way we would not have grown to such a great species that we are.

Nobody can defeat you just because you share some knowledge or science. You are called to teach and edify others in research. It might be that your junior is really fast in science and you are not feeling up to teaching him something you have learnt through lot of trials. But rest assured, he is going to find it in time and faster too. But if you teach him what you know, you will not only help him and become popular with him, you will also know your subject stronger.

You never lose by teaching and sharing skills. I know of a senior who was so apprehensive about a newcomer to research that he never taught the junior anything beyond what was immediately required for the experiments. If the junior asked any question relating to a protocol or an experiment, "just do what is told", came the reply. Today after almost a decade the senior has left research for a non-scientific career while the junior has performed well in research.

Sharing never robs you of your assets. You are your asset and not what you know.

Chapter 41

Always go neat

Oh the joy of going neat to work!

This is a small discipline that a research student can follow easily and regularly. Though this has no direct connection with your research acumen or performance, this does have an impact on your interactions, which would ultimately reflect on your research. No amount of work or research can justify going to the workplace unkempt. It is of paramount importance that we present ourselves neatly in our research environment. This is as important in research as in any other vocation.

You never know who you will be meeting on a particular day. Suddenly you might have to share a luncheon with some unexpected guests in the department. And apart from these going neat will give you a confidence that will linger throughout the day.

Chapter 42

Machines are our friends

This is one other important factor that any researcher has to remember. We all work with gadgets; research involves a host of equipment and electronics. It then becomes of huge importance that we show responsibility in our use of that equipment. They are the ones that give us the needed data; we better treat them cautiously.

For example in a biological sciences laboratory a temperature controlled centrifuge or a thermal cycler must be handled in the recommended way. It is not uncommon that in a rush, students sometimes handle these instruments roughly and that can result in a change in the calibration, sensitivity or effectiveness of these instruments. This could even affect the whole set of scientific data that is generated. Sometimes use of corrupt pen drives for data transfer can result in a complete crash of equipment. So care must be taken while handling research equipment.

Chapter 43

Back up your data

This is seemingly an elementary advice in a world of techno-buzz. But generally many of us, as researchers, often fail to back up our data regularly. I have known instances such as these. There was an instance when a student's computer along with the hard disc crashed for no reason rendering the scientific data corrupt and irretrievable. The student luckily had a significant portion of the data backed up already. But still he had to work had to make up for the loss of some precious information.

So it would be wise for you to have an external hard drive to back your data periodically. If you think a weekly back up would be fine then go for it. If your work involves daily updates then even a daily back up would be worthwhile; after all it is your data.

Of course, it would be pressing and ritualistic at times, but it will save you a lot of pain and confusion, especially when you start writing your thesis or work towards publishing your data. So it would be a worthy investment to get an external data drive or disk only for your research work.

Chapter 44

Banish insecurity

To banish insecurity is to have no paralyzing fear about the future. This might seem a lofty idea to some. Every researcher experiences times when he thinks what would become of his future. There are times when it might seem frightening to look forward into the future because of many factors like the present economic turmoil, increase in competition etc.

But I want to re-assure you; there was never an easier time in research and there was never a tougher time. One way or the other it all adds up the same way in research. Some might object to what is said here. All I want to say is: do not even spend a second 'worrying' about your future. Of course, you should have ideas and plans, but worry – no, no, no!

You will come out well and in flying colours. So keep going!

Chapter 45

Make your presentations crisp

Every research student wants to make crisp and clear presentations. In research you will be expected to make lot of presentations and talks. It is of high importance that you take every presentation seriously. I know that is tough because a presentation for the defence of your PhD thesis would be very different from a weekly presentation for your lab members. But this is exactly the place were great researchers differ from the rest. They are equally committed to making 'every' presentation good. So it is of paramount importance that one understands his subject of talk well before going for a presentation. Apart from the scientific content your technique has a strong bearing on your effectiveness as a presenter.

There are few points that are suggested for making an effective presentation. If you are using a presentation module (for e.g. Microsoft Windows PowerPoint) make sure the colours you use are neither dull nor flashy. Avoid fluorescent orange and the likes. Too many colours in your slide can be distracting.

It is important not to fill the whole slide with sentences. An unsaid rule is that a single slide should not have more than 6 sentences.

Use of images can help you make an effective presentation. It is usually the first two slides that determine who in the audience is going to be interested, so try to keep your first two slides focused, crisp and engaging. If you are presenting on cancer it would be nice to keep the first two slides on how severe the disease is by giving some numbers. Avoid animations (like a box flying from one corner and an arrow sweeping from another).

While talking it is good to avoid the usage of 'Uh', 'Mm' etc. These can be a deterrent for the audience. Overall, your preparation and presentation would decide your effectiveness.

Chapter 46

Know the players

This is crucial for a researcher! When you say you are working on a topic of research you are expected to know the other labs/teams that are working on your area of interest. You should make a list of such groups or labs. Making such a list would give you the strong stimulus and reminder that you too have a responsibility to fill up in your area of work.

Moreover, if you have such a list in your hand, you can always keep an update on what others in your field are up to, which will keep you on your toes to keep up your pace because you do not want to be uninformed. I personally had a small list of people working on my area nationally and internationally; this was very helpful while writing my PhD thesis and searching for their publications.

Knowing the big players alongside will also be vitally useful when you are writing a review for publication. Such a list would also help you in your search for post doctoral positions or jobs.

Chapter 47

Talk to your seniors

This is one lesson that I learned the tough way. I was under the impression that seniors would not like the juniors to come to them every now and then with their petty questions. I always tried to manage the show myself. But soon I realized that I cannot know everything that is needed to do my work myself and that I have to certainly talk to my seniors regarding my work. Then I decided to open myself and go to them with my queries. It was then that I realized how easy it was to ask and how willing people around us can be (provided we ask right).

Let me assure, you would get enough information if you would openly and courteously talk to your seniors. They have gone through what you are going through. They know the route. Do not hesitate to reach out and ask for help. Do not think, 'what if they don't help me?' Even if there are scattered instances where a junior was not helped by seniors, by the law of averages you will get enough positive responses to make such an effort worthwhile.

There is also one thing that a research student has to remember; we cannot expect our seniors or superiors to come to us offering help. It is the responsibility of the student who needs help to go seeking the wisdom and advice from those experienced people. There is nothing to lose by asking. The worst thing that can possibly happen is an outright denial, in which case you can always find another avenue to get your queries clarified. But the easiest and fastest way to get help and guidance is to just go to your seniors and ask. Try this the next time you are having a problem with your research and you will be amazed that people are actually more willing to help you than you think they are.

Chapter 48

Pick up the first signal for a manuscript

Oh, Manuscript writing! This is one of the not-so-loved jobs in research for many students. I know of students who would put in hours of work in the laboratory, but when it comes to writing up the work they have a sense of hesitation.

This is understandable and I have experienced such laxity in writing manuscripts for publishing. I have known people who have been sitting on a writing assignment for years. The boss would have asked them to write it down for publication but the student does not yearn for a publication or just does not start writing his findings. Sometimes the student does not find his results fit enough to be put in the form of a manuscript. Sometimes it can just be fear. Unless this fear is overcome, it would be very difficult to train yourself in this trade.

In today's research situation publication is so important and is being stressed everywhere. So keep your eyes and ears wide open; look for cues from your supervisor. When he asks you to start working a manuscript, grab the opportunity. Sit down on it and start writing, it would certainly pay off one day. Another important aspect is this: only when you start writing will you know what is lacking in your study or the loose ends that need to be tied up before you wind up your work. So always pick up the first signal and get on to writing. Failing to do this can be deleterious in the long run. Remember you have to finish the job before the heat goes off.

Chapter 49

Nobody likes a self-proclaimed genius

The title says it all. You cannot be an effective researcher if you keep boasting and strutting around because of what you do. Nobody likes it. Period!

You might be working on the cosmological basis for the universe or genome manipulation; but no one will like it if you go chest thumping because you are working on a hot topic. It is good for a researcher to be humble in outlook (even though personally he has a high opinion about himself or his work).

So even when you talk, be it science or otherwise, it would be nice to have a simple tone of conversation and not cocky. Sometimes people can even be threatened by your extra zeal (which might be mistaken for arrogance). You should know better than to sound like a self proclaimed genius. I know people who were rubbed the wrong way just because a new comer was zealous about science. Now, I am not saying that you should not be enthusiastic about your work. All I want to say is that you should be able to decide when to show your enthusiasm overtly and when keep it concealed.

Sometimes it is best to keep your enthusiasm under your skin, keeping your interests running beneath the ground and yet on the outside keeping a toned-down outlook. It's something like the duck that seems calm above the water and yet paddling strongly beneath the waters.

Chapter 50

It takes time to get trusted

If you are joining a new lab or have recently joined a lab and are pretty much the junior in there, it would be useful to know this - it takes time to get trusted. It takes time to be accepted in the team. It takes time to merge into the crowd. Just because you find that the team you have joined is apathetic towards you does not mean your situation will remain that way all through your research.

Though this need not be true always; in many situations a junior or new comer is monitored closely. So immediately after joining a lab do not think 'Oh, these people do not like me being here' or 'How am I ever going to get along with these bunch of tough people' or 'I don't think I fit in here'.

No, it generally takes time. Situations will eventually change!

But after that phase you will realize that your colleagues are not so aloof after all, and it is all part of the learning curve. So give yourself ample time to carve out your niche. Keep working on your research and people skills. One day you might even find your superiors depending on you for some of their interests and that is a joyful day to look forward to.

Chapter 51

Treat your juniors

One day you will be a senior and have many juniors working under you. At such a phase you should know not to be bossing and condescending towards them. When they need your help or suggestion there is no point in making them go around in circles just because you want to keep them dependant.

Help them, remembering the days when you entered the lab as a junior not knowing much. If you think you were not treated so nicely, you have all the more reason to be kind to your juniors. You have to break the chain. Treat your juniors kindly and they will respect you for life (and might even keep you as an inspiration). Getting the admiration and respect from juniors is not something one gets just by churning out scores of research articles in a couple of years or by showing the superior knowledge and skill in research.

It comes from the way we treat them with the understanding that they are also on the learning curve. Again as I have said before the world of research is small and you will suddenly be in interaction with one of your own juniors for a future collaboration or assistance. You might not want them to shun you just because you did not treat them right when they were working with you as a junior researcher.

Chapter 52

Extend help even beyond your lab

Research cannot progress without help from others. It is always a two way road; you give and take help all the way through research. Sometimes as students in research we might feel tempted to think that we should only be concerned about the well being of our lab and that the adjacent labs are not of our interest. Many tend to extend a helping hand only to people from their own lab.

This is an attitude that should be avoided. Whenever you can be of help to any person even beyond your own lab, do not hesitate to help unless it would be taxing on your time and research. If a colleague from another lab walks in to your lab and asks for some reagent that you have in excess, generously share (of course replacement strategy has to be agreed upon).

By helping others you only increase your chances of success at research because this is one vocation which requires huge social skills and people management. It is not only management; it is about 'being a people person'.

So let your helping hand be long and far reaching!

Chapter 53

Keep your passport ready

This was one thing that seemed so inconsequential and unnecessary to me earlier. I always thought such paper work could be done when a necessity arose, but not now. It is important that as a research student you have your passport ready even as you are working.

A chance for you to travel overseas, either for a training or conference, can arise any time and you should not be scurrying to get your passport then. If you boss calls you in to his cabin one fine day and checks with you for a sudden travel plan, you should not be hesitating for the lack of a passport; that would look awkward.

So it is important that you keep your passport ready. If you have not applied for your passport, please apply soon. Another advantage of having a passport is that mentally you are already geared up for such an overseas opportunity and being ready will help you be prepared as well. You will also be in a lookout for such opportunities. Look at it from any angle; having a passport in hand as you begin your research is good.

Chapter 54

Make it a goal to save 20% of your stipend

Ha, now we are coming to an intense issue for research students. Many research students opine that they are being underpaid. I have rarely come across a research student who felt that the stipend or scholarship available for him was enough and good. To top this all, we have the economic meltdown and other financial challenges queuing up. So why should someone take a chance with increasing economic fluctuations?

Resolve that from today you will make a commitment to save 20% of your income be it a salary or stipend. Make a dogged move and get this plan going. If you start saving 20% of your income every month, how much bulkier would you bank balance be at the end of a five year research period. You would have enough money to go without support for one full year. This period of financial sufficiency will be worthwhile as you search for post doctoral positions or jobs.

Saving money will also help you with any unforeseen expenses. You might be required to visit some prospective labs and you need not wait or depend on somebody for your travel. So cut down on unnecessary expenses and get wealthy. Saving money certainly makes life in research easy. If you think you should make the margin to 30% or 10% its fine, but let it at least be 10%.

Chapter 55

Talk to your friends and family

Being in research can be taxing and draining your energy. It is important that you keep yourself up and energetic. One of the ways in which a research student can overcome the ebbing energy during a day is through the words of people who hold him in esteem. Words of encouragement and support from such people can lift us up from the deepest pits of tiredness and lack of enthusiasm. Thus it is very important that as a researcher you talk to your loved ones and friends often. This would certainly increase you levels of energy beyond what you think is possible.

Let not your research take up so much of your time and energy that there is none left for the important people in your life. Of course they are the ones who give you a sense of belonging. Calling a friend after a tiring day can change your mood upside down and put you in high spirits. A talk with you parent or spouse can be boosting.

Try this for yourself.

Chapter 56

It's not wrong to be different

Let us get it straight: Being different is not wrong unless it is harmful to you or others. There is a tendency among researchers that a student who does not seem to gel with the crowd is acting alien. People can be put off when a person is different from the rest. Here is a truth: a person can be genuinely different, and trying to change himself for the crowd is one of the unneeded things that could be done. There is no need to change just because you are different from the crowd, unless your present condition is hampering your progress or doing damage to the general decorum of the lab/department.

For example, if you have a habit of taking a 15 minute power nap after lunch and nobody in the lab does that, there is no legitimate reason to change yourself. You can still maintain that habit which increases your efficiency and sets your mind for work the latter half of the day. Just because you are different does not mean you are unfruitful.

It would also be good to bear in mind that at times some of the not so useful habits can come in the way of effective management of your time in research. So remember it is not wrong to be different, but you can think of modifying certain aspects that can affect negatively your research and progress.

Chapter 57

Give credits

Oh no, we are not dealing with money now (I know you are relieved now, ha, ha). Knowing how to inspire mutual cooperation from our colleagues and lab members is a great skill that is important in research. One of the long lasting ways in which to develop this is by acknowledging other people's contribution to our work. Everybody wants to be felt and recognized as important. You can also increase your effectiveness as a researcher by rightfully giving due credits to others.

This can be acknowledging a small help done by a junior or a major assistance in standardizing a methodology by a colleague. Suppose your senior has helped you in an experiment and later you are having a discussion with your boss, it would be wonderful to tell him that your senior helped you. It would improve your boss's impression about your desire to be a team player. If you are in a weekly lab meeting, mentioning the help rendered by your colleagues would reflect in better cooperation next time. Just try this and see the result for yourself. You lose nothing by giving credits; you always gain.

Chapter 58

Keep abreast of politics

This is one thing that some research students do not consider important. I know this can be taken to the extreme, so a word of caution: keep this in moderation.

Political situation plays a big role in a research student's life. The world of politics undoubtedly has its effect on funding and research policies. A simple example is the difference in funding scenario for stem cell research, between a government by the Republicans and the Democrats in the United States. There are huge implications of politics in research funding and opportunities.

It is important that the student keeps himself aware of the happening stuff in politics all over the world. Sometimes it does come very handy in deciding your future labs or countries where you want to work, because you would obviously want to go to a lab in a country where there is good funding for research. So without ambiguity this is one important thing to keep an eye on. It would just take 20 to 30 minutes a day to scan the news from some of the major agencies. This way you will also be knowledgeable about the world in which you live. Knowledge of politics is important for a researcher!

Chapter 59

Failures will come your way

My friend, if there are things in life that are not going to be a bed of roses, then research is among the toppers in the list. Ask any researcher, if you need further corroboration. Research is filled with ups and downs.

While undertaking research anyone is certain to meet with times of failures. Your experiments might not work. Your manuscripts can be rejected. Your research abstract might not win recognition. Anything discouraging can happen. You can have a whole season of 'lows'. As a research student you need to remember to anticipate them (I am not asking you to be pessimistic) so you can be prepared.

The motto of the Scouts is 'Be Prepared'. This can actually be your personal motto of research as well. So always have a Plan-B for any kind of endeavour in research. This will come handy when something unexpected comes your way. Be prepared both in your scientific content and technique! You can then have an upper hand in times of failures and frustrations.

Chapter 60

Keep an inventory

This one small exercise will save a lot of your time. This is crucial for anybody in any country doing research. When you do your research, make a list of all the chemicals, reagents and equipment you use. If you are using a computer then it would be very useful to create a list of all these items and their numbers in stock.

You can systematically update the list and the numbers, as you use the items. Though this seems painful, this would save you a lot of effort. If you could combine the catalogue numbers along with the supplier's details, this can make your job of re-ordering some items pretty easy. Remember, ordering items and managing the in-use items are integral parts of research. If you can keep this part of your work easy and planned, you can focus your constructive energy and time on your work.

Chapter 61

Safety first

Yes, this might seem very plain and unnecessary. Someone might say, 'Do you think I don't know this?' No, Pal. That is not the intention.

Safety while doing experiments is important, especially when working in a laboratory with harmful reagents and potentially injuring equipment. It is not uncommon that in a hurry to finish experiments a student fails to take the needed precautions in research. Never take safety precautions casually.

I am sure you would have heard of laboratory accidents that cost a lot to research students. I know of instances where mishandling of corrosive liquids had resulted in personal damage as well as loss to the supervisor. If a particular bottle of liquid requires to be handled in a vented room, it has to be handled the same way. If there is a proper way of discarding used items, following the proper protocol is of utmost significance.

I have burnt my hands with corrosive liquids because someone left a little of them on the working table. Your life is far more important than your research or the degree you are aiming at. So keep your personal safety first. You not only take care of yourself, but also other around you.

Chapter 62

Bargaining is good; but not cheap always

It is not difficult to find research students who will talk with salespeople for a long time and communicate with them over days and weeks over items that cost small money. Unfortunately, they are under the impression that they do their labs great favours by tightening the bargaining ring around the sales people. To be noted is one important thing: bargaining as important as it is cannot overshadow the need for economical use of time.

So one does not need to really keep working at a bargain for a long time and lose precious time of research over this process. If the total money is really huge and a good bargain would help you order some other reagent or chemical, then sticking with the bargaining business is worthwhile. On the other hand, if you think there is no need for such an extended exercise, don't waste your time on measly bargains.

A costly bargain is one where you save a lot of money but end up spending a lot of your valuable time. You have to decide on these two: bargain Vs time.

Chapter 63

Your lab is your second home

Treating your lab as your home means a great deal of effort from your side. You have to be pro-active as a research student. In the course of your work and stay in your department, there will be many issues that would need fixing, which might not be directly related to your research work. You have to get geared up for those as well.

You cannot just sit idle when a fan stops working or a light goes defunct in your lab. You have to be pro-active and make sure your lab runs smoothly on all the fronts. This attitude of taking care of your lab is vital because only a sense of ownership can catapult you into the ladder of success because research is not just your data and results. Research is taking leadership, stewardship and ownership of your work and work place.

There is one great advantage in this. When your supervisor comes to know that you really (genuinely and not faking) are concerned about the smooth functioning of your lab, he will put you in charge of many important things and also elevate you faster. So it is up to you to decide if you want to be part of the audience or be an active player in your lab.

But you should also have the better sense to not make a nuisance of yourself by trying to mastermind every single detail in your lab. Thus you should have a healthy belonging and affinity towards your lab; you should be ready to take extra pains to see that your lab works finely.

Chapter 64

Humour in lab

Here we come to an aspect of working in a research environment that might be enthralling: Humour. Humour is an integral part of human life, though some people seem to defy this; big time! (Ha, Ha) You can find them morose and glum most of the time. I know of some who think they belong to the lineage of Hercules; they always seem to have that serious frown as if the whole world rests on their shoulders. They will be sighing and panting most of the time pointing out the many 'great' things they have to do.

You tell them a joke and you can see your joke dying an ignominious death as they ask, 'what happened next?'

Hey, research is not something done by dull and jaded spirits. You should have fun and humour in lab. If you are the kind of person described earlier, then all you need is a dose of good jokes. Get a book of jokes and start laughing. Laughing has its own therapeutic value (you know it).

And now you should have the common sense to not make a joke on someone in the lab (repercussions would be caustic, ha ha ha!). Make sure you have good and clean humour in lab. This will lighten your sense of burden and increase your mental alertness while working.

Ultimately it is your place of work which gives you a sense of purpose; not your graveyard. Laugh your way through research!

Chapter 65

Work even for non-thesis data

Some people are so particular and rigid that they whine and cry when asked to do some work that would not eventually be part of their thesis. They either question the need for their participation or silently resent the work and do a mediocre job. They go to their colleagues, crib on the kind of work they are doing and that they would not directly benefit out of it.

A successful researcher understands that not all the work he does will get into his thesis!

Sometimes a chunk of the work you undertake might never get into your thesis. Nevertheless, long-sighted researchers look at such opportunities to be of constructive purpose in their sphere of work, rather than hate them. There are some unsaid advantages of taking part in such work.

The most obvious one is that you will get some new knowledge of a field which might be of use at a later stage in your life. I have been part of some research work that never came into my thesis, but those were the times when I learnt some important science which later came to my aid when I started working on my thesis related research.

Secondly, you might get some good contacts as in the case of an inter-departmental collaborative work. You might even get some influential referees and favours. So be happy that people trust you with work; you are privileged.

Thirdly, you would be valued by your department and mentor for not being selfishly concerned only about your thesis. This will have positive ripples in terms of newer opportunities, favours and support.

Nothing is a waste my friend. You never lose anything by doing some worthy work.

Chapter 66

Find your T-max

Not everybody is up and running at 8 AM in the morning; by that time some would already be charting out their plans for the day and some would still be waiting for the conclusion of their dreams. Everybody has different time zones in a day.

Everybody has their specific times during a day when they will be at the peak of their efficiency and energy. It is this time when a person does his work with such fluidity and smoothness that he seems to accomplish more than any other time during the day.

You have to identify your time. My time is late in the evening. It is so easy to work and be productive when you are naturally at your best. You would find that your experiments would be easier if you plan them during this time if possible.

This does not mean you don't work during the rest of the day. You just have to make sure this time of highest efficiency does not go unutilized. Exploit this time and you will find your performance rocketing higher. Trust me; some of my best experiments were those done during such times. This is what I call T-max; the time when a person is of the maximum efficiency and energy. Your T-max might be mornings, afternoon or evenings. Find your T-max and increase your productivity. It works!

Chapter 67

Leave today with tomorrow in your hand

They say we can never possess any other day other than today. Yes, that is true. But you can still rope your bet for tomorrow. This is true in the case of a researcher.

One of the best ways to make the best use of tomorrow is to make a rough plan for tomorrow today. When you leave your lab everyday it would be great to sit down for 5-10 minutes and jot down the things to be done tomorrow. Making this list would gear you up mentally for the next day and also help you understand what kind of a day it is going to be so you can appropriate your tasks as needed.

This list can include anything that you wish to do the next day. You can include your meetings, experiments, talks etc. This will also help you to organize the requirements for your experiments if possible for e.g. you can keep your reagents for thawing overnight in an incubator if you are a biological researcher. If you are physical sciences researcher you could possibly get your equipments ready and book your time slots in advance. Making this list will give you clarity and confidence on your work plan, and will help you prioritize your work.

So never leave your lab without tomorrow in hand.

Caution: Do not bring in tomorrow's worries today!

Chapter 68

Improve your scientific communication

Good communication is a key for being a good researcher. It would be difficult to be effective in research if a student is lacking in communication capabilities. One of the aspects of communication is spoken and written language. English is the widely used language in research today. So it is indispensable for you to know this language, especially if it is not your native tongue.

I know many good scientists who are hard working and yet find it difficult to express their science in paper or before an audience. It is not surprising that they find it difficult to be effective communicators of their research. If you think you need to improve your hold on the English language you have multiple options.

One simple way is to pick up a book (need not be scientific or research oriented) along with a dictionary and start reading. This would certainly improve your language from the way you frame your sentences to your vocabulary. Another way to improve your spoken English is to enrol yourself in speaking clubs like the Toastmasters Club which would improve the way you use your language. These not only help you in your verbal language but also in your body language. And then there are also science and communication workshops arranged by many organizations. You could register for one such workshop for a reasonable amount of fees and make use of it.

Make sure you take your communication seriously. Your communication skill will have a strong bearing on your research prospects and success. But it is easy to develop this skill so nobody needs to worry much on these lines.

Chapter 69

Reward yourself

Since research is a field where there is a continuous need for maintaining momentum, keeping yourself motivated is too important to ignore. Only motivated people keep pressing on, and go forward.

It is not uncommon to find people so bogged down by the amount of things to be done that they end up lacking the motivation and the drive for research. But this need not be the case for anyone. There is one easy way to keep you motivated and driven.

Rewards!

Everybody wants to be rewarded. Rewards give us a push, a boost and a stimulus. So it would be wonderful that you to get rewards. Who will reward you and how? You need not wait for others to reward you buddy; you can reward yourself.

Upon achievements, big or small, you can reward yourself. There are many kinds of rewards you can give yourself. Suppose you did your journal club well you can take your close friend out for a dinner. May be your manuscript got published; you can go for that sci-fi movie in the nearby theatre. You can find your own healthy ways of rewarding yourself. This will keep you charged and motivated. You will feel good and develop a healthy pride.

Try this the next time you make any significant accomplishment!

Chapter 70

Start a diary

If you are my type of person then writing a diary would seem a Himalayan task; especially doing it almost every day. But this small habit would be a wonderful thing for you when you look back at your research life five or ten years later. You can record the exhilarating moments as well as the challenging ones. You can record the winning times as well as those of disappointment. You can record your own feelings and thoughts about your progress.

This would also serve as a vent for your feelings of disappointment and frustration which every researcher will experience at times. You can even include details of your day, from the experiment you did to the kind of methodology you used. You can record your discussions with your boss. You can even record his admonishing and appreciation. This exercise will keep you level headed because writing is one medium which can help you relieve yourself of any thoughts of stress and uncertainty. You can also write about your dreams and aspirations for your future and keep this as a check list or a motivating factor. If you have learnt something about your research or even life as a whole, you can record it down.

I have recorded many scientific and even life principles that I learnt during my research stint. Now when I look at them (months and years later), they serve to inspire the same courage and determination with which I had crossed many challenges in the past. So even if I sit down today feeling weak and lost, a read of my past victories do give me strength and motivation to press on.

So it is good to maintain a diary in research. Your own words can sometimes be therapeutic!

Chapter 71

Learn a new language

This is one nice thing that you could do as a researcher. This will add value to your curriculum vitae. You can learn a new language like German, French, Spanish or anything else that you would like to learn. It would also boost your confidence because as a person your talents increase in different skills; Language is certainly one.

This could also throw open newer opportunities which were not available previously. Many countries, where English is not the national language, might offer you research positions in the future, and the knowledge of an additional language might be advantageous for making a move there.

So if there is a place nearby where you can learn a new language, enrol and make benefit of it. Even if you do not use the language professionally, you would still be proud of having spent a purpose driven schedule during the period.

Chapter 72

Remember publications by the group

Have you ever seen a research student quote a work during a talk, only to be asked by someone from the audience, 'where and who?' and the student does not know the details and gets fidgety.

In such situations, if a student gives the details, it would be stellar. On the other hand if he says, 'I think it is by...mmm...' it would create a not so positive impression. People in research always like to have proper references for the works quoted; be it verbal or written. So it is imperative that a research student knows what is it that he is quoting and from where.

As a research student it would do you immense good to have a good memory of the key research papers that you are quoting while writing or presenting your research data. This would also give the positive impression about you as being aware of not only the science but also the teams that are pitting alongside, which is very important for making a credible statement in research.

Suppose you make a statement and are asked the about people who published, and you say, 'It was done by the group led by Dr.X from the University of Y' you would be impressive. So always make sure you have information about the key references that you are quoting while making a presentation or doing a write up in research.

Chapter 73

Express your thoughts with your mentor

WOW! This one strikes a chord, isn't it?

Your mentor is going to play one the biggest parts in your research. Your mentor needs to know what you are going through as part of your research (and legitimately so). Every student can go through times of toughness and stress; it would be a good thing to let your mentor know of your challenges, if you think you could get some help and suggestions.

Suppose you have a hard time with your experiments and you are feeling low and depressed, it would be wonderful to go to your boss and express your challenges and feelings than putting up a feigning confidence in the front and suffering at the back. If you think you are scared or doubtful, go and share with your mentor.

Any good mentor will understand the student and be willing to extend a helping hand. And moreover a good mentor will always appreciate your honesty and genuineness. So be sure that you always keep your boss updated not only in your science but also about your inadequacies and challenges.

Don't worry about the response. You will ultimately benefit.

Chapter 74

Post-doctoral position is not Holy Grail

Now, this is a daring statement to make during a time when a post-doctoral stint is generally the recommended norm. I am in no way underestimating the importance of a post-doctoral position for a PhD student. A post-doctoral stint for a couple or more years would undoubtedly look fabulous on your curriculum vitae and certainly be of value in your future pursuits, especially if you are planning to get into hard core research or academia. But post-doc positions also should be held in the proper perspective.

As the title of this chapter says, a post doctoral position is not Holy Grail; this has dual meaning.

Firstly, post-doctoral positions are not rare and a student getting them is fairly commonplace. It requires diligence and focus. You have to decide where you want to go and which lab you want to join. Apart from your research acumen you will also require some good references, which goes without saying.

Secondly, a post doctoral position is not the only way to succeed in research. Your success fairly depends on your commitment to your work, more than a post doctoral study. To find successful people in research with only a PhD is also not tough. There are also those who do post-doctoral studies and yet find it difficult to find a stable placement. So a post-doctoral period *per se* cannot assure you of a placement, though it can enhance your chances in some places.

So bear in mind that you are primarily the factor for your success and not the presence or absence of a post doctoral period in your career.

Chapter 75

Pay attention to details

This is a positive trait which would not only enhance your performance but also your scientific perspective. Paying attention to details is taking nothing for granted, but always seeking and looking out for an explanation or confirmation of events or pre-requisites in your research.

This simple yet profound discipline will boost your research beyond the average level. You should always be aware of what is happening with respect to your research and keep a strong vigil on your experiments. You should have an eye for picking out inconsistencies and mistakes in every protocol that you follow. Keep your eyes and ears sharp while performing your work.

For example in biology small things like temperature, acidity, balance, humidity etc can have a great effect on your overall experiment. Thus it would be of immense importance to have a regular monitor on such variables. Situations when details are overlooked are not rare; once I saw a research student doing an animal study on a rat. He was doing saline infusion into the intestine of the animal and the temperature of the liquid that was being pumped into the animal's body was 50 degree Celsius which was too hot for the animal body. When pointed out, he immediately gasped because he knew the results out of that experiment are going to be skewed and unreliable.

So it is always good to check and re-check the finer details while you work. Many a time I have burnt my fingers because of not paying attention to minute details. This happens to everyone but the crux is making a commitment to reduce this to the bare minimum by increasing your attention on details.

Chapter 76

Clarity at the end

Questions about the future are among those important things that every researcher needs to address one day or the other. May be you are in the initial phase of your research or in the middle or nearing the end. You might be contemplating this point: 'what will I do after this?' You might also be planning for many options. You might be thinking of some place to go after your PhD. If you think you know your road map, it is fine.

If you think that you are not very clear about your road map, I would like to tell you not to fret because sometimes a clear answer will come only when you reach the finish line. Yes! This might be an idea hard to swallow for some, but it is true as many would attest. So the best thing to do is to just keep planning about your future without panicking. Keep your options open but do not fret if you don't seem to have a clearly charted plan.

Just because a student does not have clarity of future (right now) does not mean he will not be successful. There have been many great achievers who did not have a clear blue print about what they wanted to do when they were young, but as time went by they slowly got into the rhythm of their career and began letting out their wings to soar high.

So an absolute 'A-to-Z' clarity of your career path is not an imperative in research. Sometimes you have to reach the end of the fog to see clearly what lies ahead.

Chapter 77

Slip of the tongue

Uh-huh! Tongue and its powers!

There is a tough observation in research: talented students losing support of their peers and superiors because of a loose tongue. Having a loose tongue that keeps prattling and letting out unneeded words can be a disaster in research. If a student does not know how to control his tongue he is bound to hit rocky waters soon. A loose tongue does not necessarily mean a maligning one. It can just be the habit of blurting out statements without giving any thought to its impact on the hearer. A seemingly harmless comment can actually expose you to the ire of another person. So prudence has to be exercised in using your words carefully.

I have burnt my fingers because of my tongue. During my early days in research, I was pretty naive and spoke my mind out without weighing the repercussions it would engender. But soon I realized that my uncontrolled talk was a big factor sabotaging my own growth and success. It was then that I made a commitment to myself; never to say anything without thinking it thoroughly through.

I know this person with a bright mind for science, who because of loose usage of words has incurred losses in the workplace. That person's reputation was maligned because of unwise use of words. There was such a loss of precious image and time for that person. It was just a small indiscipline with tongue that really took a heavy toll on that person's career. This guy let out a seemingly benign opinion about a sect of people in a casual talk (remember 'casual' talk) only to realize later that one of his audience belonged to that particular group. That one colleague was so offended, that he created such a nuisance for this guy in the next few years. And this person regretted opening the mouth and letting the tongue loose.

So it is important to use your tongue wisely in research. You can be caught unaware by your own tongue. Never...never...never let any un-meditated word come out of your mouth even if it means that you have to think for a minute or two before opening your mouth. Always be cautious with your words.

Words are precious and they can prove costly if used unwisely!

Chapter 78

Pat your back and walk

Research can sometimes prove lonesome. Sometimes you enter a season of dryness in your research when you just don't feel up to it every morning. If you are in one of those seasons then it will be of great importance to make sure your momentum goes unaffected. Dryness in our research, compounded by lack of appreciation and encouragement from others, can have a strong effect on our overall motivation for research and our performance.

It is during these times that you have to push yourself forward tenaciously. Such periods can be those short or extended times when nothing significant comes out of your work or no publication seems to be possible etc. Such periods are universal and every researcher will have his share of it. So do not think only you are 'divinely chosen' for this 'time of testing'.

For you to be effective in your work, you have to develop the maturity to tell yourself that it is all right to have a period of dryness and lack of productivity. It is during these times that you should remind yourself of past victories; remind yourself of yesterday's achievements to keep pushing today. You should be able to pat yourself for the achievements you have made already and keep moving.

Even if you do not find anything, just pat yourself for the fact that you have come this far and many have already fallen off the track settling for something else. Not quitting is a great virtue, my friend. You can really kiss yourself for not quitting so far. It is this capacity, to be your own positive coach and encourager that will take you longer in your research.

Chapter 79

Be trustworthy

Trust is something that is so rare in this world; people who can be trusted are highly valued and sought after.

This is true in any field of vocation and especially in research. It really takes a lot of courage to be a person who can be trusted. A person who can be trusted is one who does what he says he will do and speaks what he has done. He neither adds nor subtracts anything from these. Keeping yourself trustworthy will be a painful but rewarding discipline.

You should have a high standard of conduct that you adhere to, which you would never compromise for anything. Even if you incur loss because you have a high value system, that is commendable rather than being a person who swims according to seasonal gains. Be a person who can be perennially trusted. It would be painful to start being such a person but as many attest, it is worth the try.

Let it be your mission to become trustworthy. If your boss can give you a job and sleep without worrying about its execution, that is a sign of your trustworthiness. If your colleagues can count on you at all times and in all situations, that is being trustworthy.

So make it a commitment that you will be and always remain a trustworthy researcher. You will not only be valued in your own circle but your research and influence will spread.

Chapter 80

Publish or get left behind

Doing great research is awesome and so is publishing those results in a journal of good impact. People in research are aware of this phrase – 'Publish or Perish'. As morbid as this may sound, it reiterates a key truth; people who do not publish significant data slowly get into oblivion. Even your progress in the hierarchical system of your institution might depend on your publication history.

It is very true, because publications in indexed journals is one of the ways in which people generally gauge someone's research capacity and growth. It would be wonderful to always remind yourself of this fact that you have to publish your work and not stop short of it. Some people might superficially say, 'I don't really think publications are as important'. But the truth is that sometimes your publication list will speak a lot about you; more than what you could say verbally.

So make it a goal to publish a minimum of one review article and one research article relating to your thesis. Nobody prevents you from doing more; sky is the limit. I know of people who have more than fifteen papers only during their PhD period. And then there are those who have just one research publication. Though quality of your research publication is most important, numbers still matter; to keep a track of our progress.

So it would certainly be really worthwhile to have the goal to publish 'at least' one review and one research article during your PhD.

Chapter 81

Intuition helps

Sometimes what we think is wishful thinking might actually become a reality for us in research. Intuition is one such thing. Haven't you heard of scientists getting key ideas in research while they were sleeping or playing a game?

I personally know of one instance when an idea for a novel medical diagnostic company came to one of the founders during a casual game on the ground. Yes, our intuition is one of the many ways in which our mind can really nudge us forward in our research.

This is one area which can be debated. Sometimes you can be in a situation where you are not able to come out with some solution to a problem and suddenly you have a flash of idea that would fix your predicament. Intuition can sometimes help you decide some important things while doing research like choosing the right lab, institution or country. Even in your research, intuition might come as your aid. Intuition can give you ideas such as a new experimental design or a better methodology.

So whenever you get some idea or plan, make a note of it, however great or stupid it seems. Sometimes great ideas evolve from a sequence of not so great ideas. Suddenly one day you might come up with a WOW idea. Keep yourself open to your intuition! The idea to write this book came about in a flash of a second one day as I was in my lab; what you are reading right now is one strong proof of the manifestation of intuition.

Chapter 82

Whatever you decided in the past is the best

It is very common that a research student comes to a point in time where he questions his very decision to come take up research. I have heard research students whine and lament, 'Why did I choose research?' or 'My friends are all settled, earning a six figure salary and I am here struggling with my lean stipend'.

I want to tell you one encouraging statement: Whatever you have chosen (your choice to come into research) is the best thing that you have done to yourself. You might say, 'Oh is it so? I wish I could believe it'. Yes! Your decision to come into research is the apt decision that was taken by you.

Your choice of entering research would certainly have been based on many factors. They include your educational, social, familial, personal factors. Anybody in your shoes with all these factors exactly like you (your mind, your situation, your education, your society, your family etc) would have exactly decided the same thing. So you could not have decided differently and neither could anybody else in your place.

So it would do immense good to anyone stop thinking that he has made a wrong decision by coming into research. No such decisions are wrong; they are only challenging or easy. May be you have chosen a tough route but never the wrong one. So believe in your decision and choice to get into research. Certainly something good will come to you and you will realize this soon. So just keep going. One day you will be glad you did.

Chapter 83

Take care of your body

This is a piece of suggestion for a vast population of research students. And the title says most of it. Research students have lots of things to do, many papers to read, presentations to be prepared that they tend to overlook the need of monitoring their health and health related life style.

I know of research students who come and say with a smile on their faces as if they did a commendable job - 'Today I skipped my breakfast. Obviously I had to do a lot of work'. Such statements are also silently glorified by many. Some also say, 'I was feeling sleepy, but I forced myself to keep awake yesterday night'. Now apart from these, students also expend their energy in entertainment. I have been told of students who stay awake till 2 A.M. watching movies with their roommates. How then do they expect their body to deliver during the working hours? Moderation in everything is the key.

At a young age it might seem okay to put the body under such stress thinking you can compensate later. No. That never happens; either you take care of you body now or regret later. Keep a regular timing for food, rest and exercise. It will help you in the long run. It will also keep you agile and active.

Chapter 84

Be ready for surprises in research

Ha! Being in research is one of the hottest things because you never know what will come up in a day. Research always brings with it some element of unexpected surprises. Some might be pleasant and some wrenching. But you have to always keep an eye for surprises.

Research is something like the Hare and Tortoise race. Unexpected events will arise. Sometimes the tortoise will win. Then the Hare will learn from its mistake and make a winning run. Sometimes a rat will jump in to the race and take the medal.

You might be progressing very well and yet you might find your colleagues churning out publications. Sometimes you might find a newcomer outshine the rest of your group. So keep a calm countenance and keep your momentum going. Be ready to accept some surprises in your career.

Chapter 85

I am truly sorry

Like any other interaction, research requires that you develop your people skills and any student who can develop a good skill of dealing with fellow beings will undoubtedly have a vantage point. This is true because you cannot possibly take care of every nail and nut in your research; you will obviously need the help, support and good will of others. In such a process of interaction, by probability, you will (and so will anyone) make some errors or mistakes in dealing with people. Take heart; this is fairly common.

They might be your words or actions. If you think that by your words or actions you have done some damage to someone during your day, go ahead and make up for it as early as possible. It is good to break the ice as early as possible. Express your deep felt apologies for your mistakes to someone to whom you owe one. It might be your guide or superior or even your junior.

Telling them you are sorry for something does not make you lowly or below standard. It will only elevate your image and ensure that you continue to get the wonderful counsel and support of those who are pivotal for your growth. But caution has to be exercised not to make this exercise a passive, senseless and ritualistic monotony. If you are sorry make sure you say it like you mean it.

Apologizing is a key skill in any relationship and thus it is of a huge value in research. If you think you have to make an apology, do it the first thing tomorrow.

Chapter 86

Who will follow you?

Everybody wants to be leaders in their own circles. We all want to be known as a person with great leadership skills. Research requires people like you to become leaders who can lead science as well as people to greater heights. But the question is what kind of a leader you will be. Don't we all know of leaders who are so pompous and thumping? They flaunt their achievements and position, and try to exercise what they think is leadership.

Many a time, leaders do not follow the very things that they profess. In research, too there are leaders who develop standard operating protocols in lab, but fail to follow those guidelines themselves. Such a kind of leader or superior or colleague would eventually lose the impact and fail to be a positive contributing factor to the lab.

If you want to have an impact in research circle, people should find you following what you want them to follow. If you are a person who believes in punctuality it is imperative that your juniors and colleagues find you in your workplace on time. If you want others to handle the laboratory equipment and reagents judiciously, you first have to be known to follow the same. People are looking at you for what you follow than what you say. So if you are a neatness freak (as some may call it), show others at all times that you really keep yourself and your surroundings clean and neat.

You will be respected and held in esteem. This will certainly have a bearing on your command in your own circle of research. Follow what you preach!

Chapter 87

Think positive

Whoa! Everybody knows this. Then why should we repeat this?

Just for the reason that such an important factor should not be missed out when trying to list the wonderful ways to being a better researcher. If you take any hugely successful person in your area of research, you can find them positive and filled with great hopes. No negative thinker ever succeeded in research because pessimists do not find the stimulus and motivation to drive them to greater causes and pursue challenging roles in research.

There is no dearth of people who focus on the negatives around them. The pessimistic statements they make will put others off like, 'My boss, is so biased that whatever I say will not fall into listening ears' or 'Everybody around me seems to be scheming against me' or 'Why did I come into this hellish place'. Come on, nobody that pessimistic could ever hope to become successful in research or lead a future team.

Only positive and enthusiastic researchers are the need of the day. So even if somebody has been pessimistic and mouthing negative statements as often as the second hand in a clock, they can make a decision to start focusing on the positives around. A good book on positive thinking should really help change the perspective and outlook.

So to become a better researcher, developing a positive outlook towards research and life as a whole is indispensable. Keep thinking upbeat. Thoughts are the 'only' things you can really control in life!

Chapter 88

Be open to ideas

Sometimes the greatest inventions came out of simple suggestions. I was told the zip used in bags and pants were invented by a person who was suggested that he find an easier way to close and open dresses.

People not only laughed but also ridiculed the idea that man could fly with the help of a machine. But today flying has become commonplace.

So learning to receive an idea or suggestion without making a mockery of it is a valuable lesson every researcher has to learn. Nobody knows what could come out of it. If somebody suggests you something, make it an unsaid rule never to ridicule or be contemptuous. Sometimes you might actually be touching the tip of a gold mine and not know.

It would be great to show respect and be open to ideas, even if you don't try them personally. You never need to be so rigid that you tell somebody on their face that something will not work. Ideas can come from anyone. So be ready and be open to ideas.

Chapter 89

Let personal problems be personal

Under-performance in research is a common problem for many research students. One reason for such decreased performance is tendency of these students to let their personal problems affect their professional work.

Everybody in this world has problems. This can never be an excuse for not performing well in research. Of course, your problems can be tough and challenging, but being a researcher, you have to rise up to the mental maturity to draw boundaries between personal problems and professional responsibilities. It is never welcome if you are a person whose emotional quotient is so low that challenges of the personal life start spilling into the areas of professional duties.

It is not uncommon to find lots of young talent in research wasted because of personal problems encroaching into their research responsibilities. You might be facing many problems apart from your profession and it would be wise to not let them become a reason for any hiccups in your research. It is your duty to make sure that you will be responsible and goal oriented, so that you will take whatever measure is required to make sure your research goes unhampered, even in the midst of personal pressures.

As you enter your working hour every day, it would be great to decide that you will be doing your duties with utmost commitment and responsibility, and not let your other areas of life affect your productivity. This trait is one of the top jewels a researcher can possess.

Chapter 90

Value life

A successful researcher does not work only on the external job requirement. He is also known for his intellectual standpoints regarding various issues of ethics. Ethics is valuing another life to the point that you will do nothing as part of your research that would degrade or trivialize another life, be it animals or humans.

Valuing life is an inseparable part in developing a healthy research ethic. These are days when human rights and animal ethics are upheld at higher levels. So if you want to be a good researcher, it is vital to understand the importance of valuing life.

Without a basic value for life and passion to improve the standard of life, you will not be able to have the strong drive to stick to your research and produce significant results. What does it mean to have ethics in research? This would mean that every single experiment or test involving humans or animal be done with proper protocols.

If you use patients or patient derived samples for your research, it is highly important to have a comprehensive consenting proforma for the patient. The patient should be informed about the work. Similarly, treating patients with respect and gentleness is among the highest priorities in medical research. If your work involves animals, then take measures to make sure there is humane treatment given to them. These are things that might generally go unnoticed by your peers but as a person, when you follow these your sense of personal worth and integrity will rise. You will certainly feel and be a good human being, which is as important as doing a great research.

Chapter 91

Not all great scientists get a Nobel

When we truly understand the purpose of research then external recognitions of achievement like Nobel Prize etc get into the right perspective. Generally people think that getting a Nobel Prize is the pinnacle of research achievement. No! That cannot be farther from the truth.

Such recognitions are just boosts and motivating factors for researchers. Almost every research student would have mentally imagined getting a Nobel Prize for his work, at least when he started his research work. Haven't you ever visualized yourself as a Nobel Laureate?

Getting a Nobel Prize is wonderful but true research is beyond such recognitions. The point of research is to make better availability of facilities and avenues to improve human life. I know of wonderful scientists who silently work on diseases like diarrhoea, which kill millions of children every year.

So, if Nobel Prize is to be considered the hallmark of a great researcher, then we are missing the point of research.

Aim to be a scientist who is committed to improving the quality of human life; a scientist who is bigger than a Nobel Prize.

Chapter 92

Never be intimidated by a talk

When I started my research career about a decade ago, I cancelled my very first scientific talk. Just minutes before I had to deliver my talk on that particular day, I went into the office of my professor and told him I wanted to cancel the talk.

I was scared and thought I would be humiliated if I went ahead with the talk. He was understandably annoyed. It was a great embarrassment for me. I did not believe in myself. I was 'intimidated' by the prospect of talking before big scientists and professors. I feared I will be ridiculed and laughed at.

But fortunately my professor was kind enough to call me later and tell me that I could either have informed earlier or could have gone on with whatever I had prepared. I learnt an important lesson that day that taking a journal club or giving a talk can always be a tough challenge; but that need not put a pressure on our blood vessels.

We have to make sure we do enough preparations, then go ahead and do the talking. Whatever will be, will be; so just go ahead and do what is required. Let the whole research community jump on you later, but do what you have to do.

Talk it out!

Chapter 93

Watch your notice board

This is a small thing that could be implemented by any research student in any lab. This could keep you updated and also well informed about many things.

Generally, many students miss important information because a significant percentage of information is normally pinned on to the notice boards of the department. I know of many students who never care to stand by the notice board to have a look at what is stuck there.

There are many advantages of having an eye on the information boards in your department. You could get information about fellowships, scholarships or workshops, which might not be broadcasted as an email to you. It is just a small discipline everyday to have a look at the department's notice board to see if there is any relevant information for you as a student. So having a check on the notice board could be a potential help for a research student.

Chapter 94

Don't grab more cups than you can drink

This is very relevant to students who are new to research. This can also be true of people who have been in research for a longer time.

Many times a student, in order to show his capability and commitment, obliges to do multiple things. If he is able to manage all the tasks he has undertaken, he is the hero. But not so infrequently do we see many students, in an attempt to satisfy their guide or superior, take up more responsibilities than they could handle, eventually resulting in strain and stress.

This is one thing that could be easily avoided. Never attempt to take so many tasks on your shoulders that you falter under the sheer weight of them. It would be wise to know your strengths and decide what you could pull off. There is no point in accepting to take a journal club, order a list of samples and visit another lab, on the same day when you also have a long experiment to finish and you have to catch your train at 4PM.

You have to plan on what you take up for a day. This would not only save your face but also help you maintain your efficiency. So never think you can take all the cups and still manage the show. It is better to exercise prudence on what you can handle and deliver.

Chapter 95

White board your ideas

We think well when we see what we think!

This is a suggestion for people who want to juggle multiple ideas and squeeze out some kind of a solution. Having a white board on the wall of your room with an erasable pen would save a lot of paper (you go green, buddy!) and also help you orchestrate your thoughts.

Imagine a day when you return from your lab mulling over a research problem. How good it would be if you could put your ideas and flow diagrams on a large board to sit back and think it through. You can strike it, re-write and strike it again. This flexibility will certainly help you bring your thoughts together into a schema. This approach, of board and pen, would really help.

You can also organize your scientific priorities and take a look at the responsibilities and plans for the coming days. Sometimes you can just jot a couple of points or ideas in one corner to revisit them later.

Certainly having a board and pen in your room would be very useful. It should not be expensive to get such an arrangement in your room. Moreover you will find it immensely useful.

Chapter 96

Statistics

Friend, there is one important thing in science: statistics. This is one area which might not be appealing to every research student, but this is certainly an area that cannot be avoided or overlooked. Statistics help us to derive meaningful inferences from our data.

When you present your data in a conference or write it up for publication, the statistical validity of your study would be very important. Thus it is very important for you learn the different statistical tools used in research. This would also help you while you write your thesis at the end of your research. It would also be very useful in the long run if you are going to establish your own lab because you cannot expect your students to update themselves with statistics unless you learn it first.

Learning statistics should be one of the important things for a research student. I know of good researchers who, for their lack of statistical knowledge wait upon some busy statistician for analyzing their hard earned data. They wait for hours and days to get an appointment with a person who is certified in statistics.

So would it not be great to learn it yourself and become more independent? Of course we are not undermining the importance of the inclusion of a certified statistician in a research project, but at the same time a research student's statistical abilities will also be very important.

Chapter 97

Software

Oh yes, software are one of the important tools in our hands today. From our waking minute to the time when we doze off we are in continuous touch with some kind of software. Be it our mobile phones or computers, our life is so wired.

This is so very true of research as well. A research student, apart from his subject, has to be literate in different software. It would certainly come handy when analyzing and compiling your data. Today software is being used for so many techniques that any research student dare not avoid them.

You would be well armed in research, if you will learn different software connected to your area of work like those used for statistics, plagiarism, language etc. If you are a person who uses different instruments for your data acquisition and analysis, then it is a worthy investment to learn the software that come along with the instruments.

A person who knows the different software used in his field will certainly be respected and sought after. Apart from being better at your own research, you can also be of help to many others.

Chapter 98

Thesis writing

This is one of the most feared parts of doing a PhD, by a good number of students.

I dreaded starting my PhD thesis. I kept pushing it day by day, just because of the fear of undertaking such a massive task. But in the end I had to finish it. But one thing that I learned as I started writing my thesis was that it all has to begin with one thing: Beginning.

Unless you begin writing your thesis you will never be able to know what it is that you are expected to write. Of course your guide and mentor would be always there to help you make your thesis better. But even your mentor cannot help you unless you start writing it. So the first and important step from your side will be to begin.

So if you have completed your research work and are nearing the deadline to submit your thesis, dare yourself to begin the writing process. It would certainly help to find a place and time where you can write without having too many disturbances as a good flow of your thoughts is crucial for writing a thesis. May be your department library would be fine. Sometimes your lab is the best place to sit and write. I wrote most of my thesis in my lab during the nights when there would be the least disturbance.

So the best way to get moving on your thesis writing is to just begin. As you start writing, the pages will fall one by one in place (trust me!). And don't forget to take the help of your mentor; you will write a great thesis.

Chapter 99

Celebrate your life

Whether you are in research or any other vocation this one factor is very important for you to stay afloat, especially when more and more responsibilities come your way. The more you grow and reach higher in the ladder, more will be demanded from you. As you progress in research you will be promoted with a different set of responsibilities.

At such a time, you should remember that as a researcher you are expected to be responsible. At the same time as a human being you should not fail to celebrate life. What is life without celebration?

Without a spirit of celebration, no amount of achievement or progress would satisfy anyone. It is important to have an outlook towards life that wants to celebrate the sunny side of life.

So go ahead and celebrate your life every day. You need not wait till you get your Nobel Prize. Start your celebration today for all the good things you have! All the best!

Epilogue

Firstly I thank you for choosing to pick this book. I thank you for your patience in reading this book. But more importantly I hope you found something interesting and useful inside. This book was written with the simple intention to give a set of guidelines to a researcher. So even if one suggestion in this book was useful and has found favour with you, I shall consider this attempt worthwhile.

If you found this book useful, would you consider recommending it to others, thereby being a part of this positive chain of action?

Encourage others, tell others about these points and share this book with them. We can together make a better tomorrow, than the yesterday that we have lived. It is in your hands and mine. So I strongly urge and encourage you to take up your research with renewed fervour and mount up with wings as eagles to soar in research.

I wish you all the very best of research. I thank from the bottom of my heart and take leave with your permission.

Do post/send in your comments and suggestions; they will be highly appreciated.

Sammy **Jarauvik**

sammyjarauvik@yahoo.com

Twitter: @sammyjarauvik

YouTube: sammyjarauvik

Other book by this author

Can I be courageous? Is patience relevant in today's world? Is there anyone I can trust? How can I regain my youthfulness? How to improve my relationships? How can I handle criticism? How can I reach my goals? Failures upon failures; now what?

Have you ever asked such questions? This motivational book is the result of such questions!

Available @ Amazon, Createspace and other portals.

www.ingramcontent.com/pod-product-compliance
Lightning Source LLC
Chambersburg PA
CBHW051810170526
45167CB00005B/1960